Making Trainers: The Supply Chain Process

Nicolas Brasch

Making Trainers: The Supply Chain Process

Text: Nicolas Brasch
Publishers: Tania Mazzeo and Eliza Webb
Series consultant: Amanda Sutera
Hands on Heads Consulting
Editor: Kate Daniel
Project editor: Annabel Smith
Designer: Leigh Ashforth
Illustrations: Sasha Kolesnik
Project designer: Danielle Maccarone
Permissions researcher: Lumina Datamatics
Production controller: Renee Tome

Acknowledgements
We would like to thank the following for permission to reproduce copyright material:

Front cover, pp. 28, 32: Zarya Maxim Alexandrovich/Shutterstock.com; pp. 1, 19: Duncan Andison/Shutterstock.com; p. 4: michaeljung/Adobe Stock Photos; p. 6: wertinio/Shutterstock.com; p. 8: Gorodenkoff/Adobe Stock Photos; p. 9 (top): KENA BETANCUR/Getty Images, (bottom): Cover Images/ZUMA Press/London/United Kingdom/Newscom; p. 10: NosorogUA/Shutterstock.com; p. 11 (top): lithiumcloud/iStock/Getty Images, (bottom): iStock.com/gorodenkoff; p. 14 (top): Max Barnum/Shutterstock.com, (bottom): Haider Y. Abdulla/Shutterstock.com; p. 15 (top): shams Faraz Amir/Adobe Stock Photos, (bottom): iStock.com/alffoto; p. 16: Pedal to the Stock/Shutterstock.com; p. 17 (top): panpote/Shutterstock.com, (bottom): Roman Lacheev/Alamy Stock Photo; p. 20: xPACIFICA/Stone/Getty Images; p. 21: iStock.com/DEBOVE SOPHIE; p. 22: Klaus Oskar Bromberg/Alamy Stock Photo; p. 23 (top right): Kingsimon/Alamy Stock Photo, (middle left): iStock.com/RichLegg, (bottom right): iStock.com/simonkr; p. 24: JONAS ROOSENS/AFP/Getty Images; p. 25: chayakorn/Adobe Stock Photos; p. 26: puhhha/Shutterstock.com; p. 27: iStock.com/Nikola Stojadinovic; p. 28 (top): maicasaa/Shutterstock.com; p. 29: Steve Marcus/REUTERS; p. 30 (top left): Gorodenkoff/Shutterstock.com, (middle left): panpote/Shutterstock.com, (bottom left): allensima/Adobe Stock Photos, (top right): iStock.com/danishkhan, (bottom right): Gorodenkoff/Shutterstock.com.

NovaStar

ISBN 978 0 17 033494 5

Cengage Learning Australia
Level 5, 80 Dorcas Street
Southbank VIC 3006 Australia
Phone: 1300 790 853
Email: aust.nelsonprimary@cengage.com

For learning solutions, visit **cengage.com.au**

Printed in Malaysia by Papercraft
1 2 3 4 5 6 7 29 28 27 26 25

Nelson acknowledges the Traditional Owners and Custodians of the lands of all First Nations Peoples. We pay respect to Elders past and present, and extend that respect to all First Nations Peoples today.

Contents

What Is a Supply Chain?

Products don't appear on shop shelves out of nowhere. They are not delivered to your home by magic. A large number of people and companies, all around the world, play a part in getting **goods** to you. These people and companies make the designs, find the **raw materials**, **manufacture**, deliver and sell a product. Together, this process is known as the "supply chain".

A shopper browses trainers in a store.

Each step in the supply chain depends on every other step. So, if there is a shortage of raw materials to make a product, the supply chain breaks down. If machines stop working in the factory that makes the product, the supply chain breaks down. If the ship carrying the product from one country to another gets held up in the **port**, the supply chain breaks down.

The supply chain for making and delivering products like trainers is made up of different steps.

Making Trainers: The Five Steps of the Supply Chain

In this book, you'll find out how trainers are made, following the supply chain from beginning to end. Trainers, runners, sneakers – there are many names for them. They are the world's most popular shoes. They are worn by athletes, factory workers, office workers, teachers, celebrities, and probably you.

Trainers come in many different styles and colours.

1 Designing Trainers

Step one involves designing the trainers. Everything needs to be designed before it can be made.

2 Sourcing Raw Materials

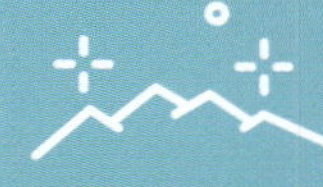

The next step is gathering the raw materials that the trainers are made from. Some of these raw materials may need to be taken out of the ground. The suppliers of these raw materials then sell them to the companies that make the trainers.

3 Manufacturing

The third step is manufacturing – the making of the product. Not every type of trainer is made the same way or from the same materials.

4 Warehousing and Logistics

Next is warehousing and **logistics**, which is where the trainers are stored and how they are transported.

5 Retail

Finally, we look at selling the trainers. **Retail** is one of the ways in which trainers find their way onto the feet of **consumers** like you.

STEP 1 Designing Trainers

What do you think is the main purpose of a pair of trainers? Is it improving athletic performance or making a fashion statement? You might think one thing, while your friend thinks the other. Or maybe you think it's both.

A shoe designer needs to think very carefully about the shoe's purpose when they design a trainer. All the **features** that are put into a trainer need to meet this purpose. Examples of purpose include extra grip, speed, comfort and good looks. So, if the designer wants the shoe to improve performance *and* be a fashion statement, then the trainer must not only have the latest scientific developments for performance, it must also look stylish.

Designers use software to develop the look of a pair of trainers.

A pair of trainers designed for a champion tennis player has special features to help the player to stop, change direction and run faster.

Clever Thinking

A British designer, Kiki Grammatopoulos, has designed a special outsole, which is the bottom of a shoe, to fit around trainers. The purpose of this outsole is to pick up and spread plant seeds while someone is running. This copies nature, where seeds are spread by sticking to animal fur and then dropping somewhere else.

Kiki Grammatopoulos's Rewild the Run outsole

The Design Process

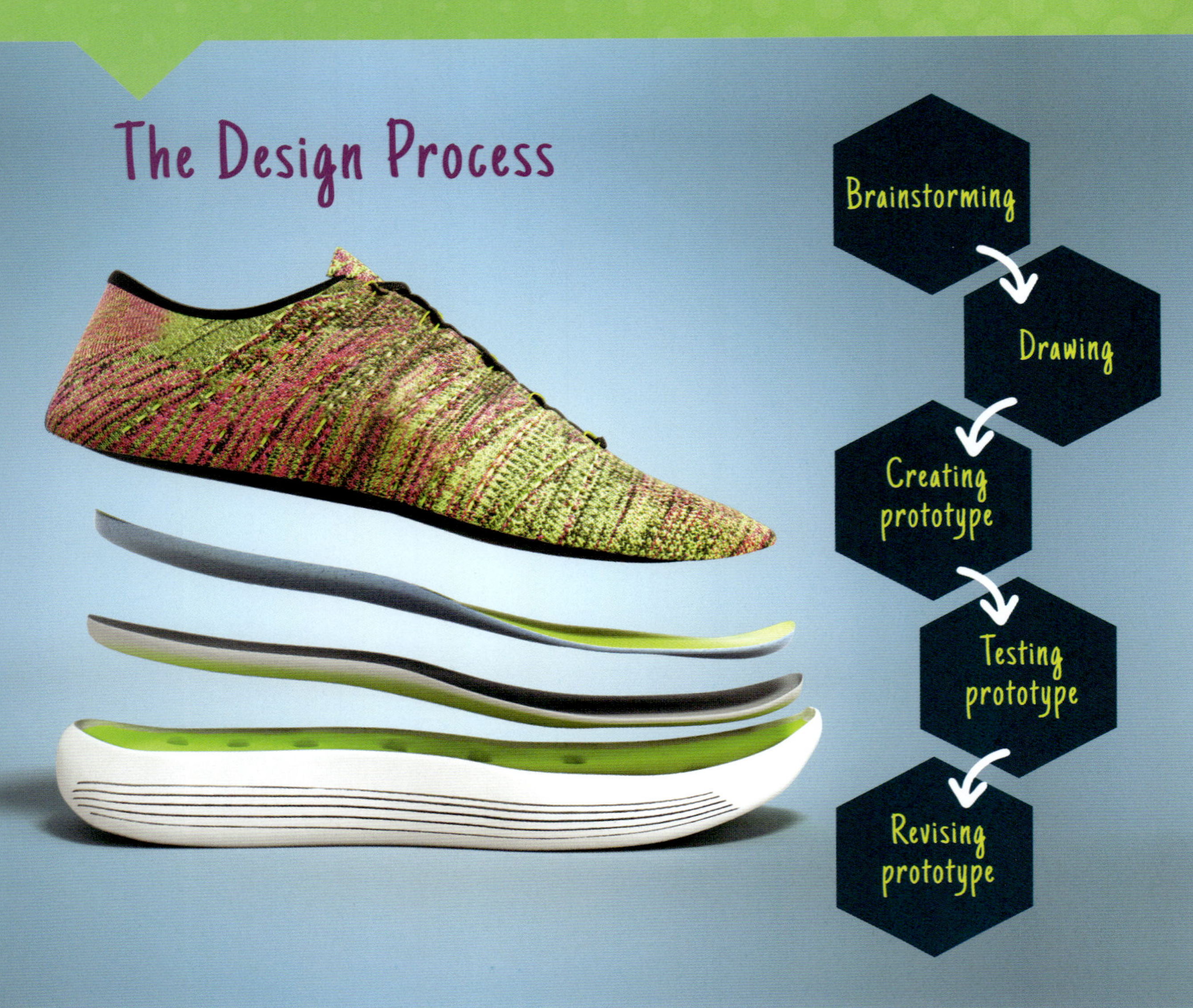

The design process for a pair of trainers starts with a brainstorming session. Brainstorming is when a group of people come up with ideas, such as which materials to use, which features to add and which colours to use, and discuss them until they decide which one is best. Often, the final choice is a combination of one or more ideas.

The next step is drawing the trainer that is going to be made. While most designers draw using three-dimensional (3D) computer technology, some still prefer to start with a pencil and paper. The drawing includes all the features that the designer wants in the type of trainers they are designing.

Once the designer is happy with what they have drawn, a prototype is made. A prototype is a model of the shoe that follows the design. A designer sometimes changes their design after seeing the prototype because it did not look as good as they intended. Then another prototype is made of the new design.

The prototype also has to be tested for performance. A lot of the testing takes place in laboratories, where scientists put the trainer through the same sort of force and pressure that it will come under in the real world. This testing often reveals failures, such as the shoe wearing out too quickly, that means the designer has to start again.

A designer works on a model of a trainer.

Become a Designer

If you are interested in a career in designing trainers or other products, then you can study design technology subjects at high school and then do a course in **industrial design** at university or college.

A product designer uses computer technology to create a detailed 3D design.

STEP 2

Sourcing Raw Materials

Raw materials are the basic materials that trainers are made from. Trainers were originally made from natural materials, like rubber, which comes from rubber trees. Today, more than forty different materials are used in making trainers, some natural and some human made.

Among the most common are:

- leather
- rubber
- EVA foam (a type of soft foam used for padding)
- polyester (a type of human-made substance that can be made into **fibres**)
- **artificial leather**
- cotton
- metals.

The metals are mainly used for the eyelets (holes for laces) and the tips of the laces. The other materials are used to make the sole and the upper part of the trainer.

Trainer Raw Materials from Around the World

Where Do the Materials Come From?

Raw materials for trainers come from farms, mines, laboratories and factories all over the world. The raw materials for trainers are gathered by the suppliers and bought by the shoe **manufacturers**, so they can make their trainers.

Most of the raw materials for trainers come from these places.

Using Sustainable Materials

Today, most manufacturers of trainers use some recycled materials and materials that are less harmful to the environment than those used in the past.

To obtain or produce materials sustainably, many manufacturers of trainers:

- Use plastic waste collected from waterways and coastlines, which creates a cleaner environment.
- Melt recycled plastic water bottles to produce polyester strands, which reduces plastic waste.
- Use recycled polyester, which reduces the amount of polyester that has to be produced in laboratories and factories and saves energy.
- Use recycled rubber and natural rubber, which reduces manufacturing **practices** that harm the environment.
- Use wood fibres instead of cotton, which is better for the environment because cotton needs a lot more energy and resources, such as water, to grow.
- Use leather from companies with environmentally friendly practices.

plastic waste

shredded rubber tyres

Merino sheep produce fine, soft wool.

Wool has been one of the most important agricultural products in Australia and New Zealand since European **colonisation**, and one of the best quality wools is merino. Merino wool is a sustainable product now being used by some manufacturers as a material in trainers.

Helping Cotton Communities

One of the raw materials commonly used by manufacturers of trainers is cotton. It is used to make the upper part of the shoe. Several large manufacturers of trainers are members of the Better Cotton **Initiative**, whose mission is "to help cotton communities survive and **thrive**, while protecting and restoring the environment".

Workers hand-pick cotton in Brazil.

STEP 3

Manufacturing

Manufacturing is the process of making something into a finished product. Once the raw materials for trainers have been obtained, they are transported to a factory where the trainers are made.

How Are Trainers Made?

Most trainers have two main parts: the upper part and the lower part (known as the insole).

There are six main stages in making trainers:

1. Mixing the raw materials: The raw materials are mixed together using various processes, such as heating, cooling, hammering, stretching and compressing, and made into huge rolls.
2. Stamping and cutting: The rolls of mixed materials are laid out, and the shapes of the upper and lower parts of the shoes are stamped into separate pieces and then cut.

materials production

3 Stitching and punching: Holes are punched in some of the pieces, including the holes for the shoelaces, and then the separate parts are stitched together. The punching and stitching is usually done by machines but in some factories it may be done by hand.

holes have been punched

4 Inserting the insole board: A shaped board, often padded, is inserted inside the shoe, over the insole.

5 Glueing and cementing: Hot glue and cement is used to bind the upper and lower parts tighter, to support the stitching.

6 Quality control: Some of the trainers are tested using special equipment that **simulates** real-life wear and tear. This identifies any problems that may occur, such as stitching coming undone, or soles not glued together properly.

a quality control check

Where Are Trainers Manufactured?

Most trainers are made in huge factories in Asia (particularly China, India, Vietnam, Indonesia, Taiwan and Bangladesh), and in Central and South American countries (such as Mexico and Brazil).

These countries are preferred by the companies making and selling trainers because they have low **labour** costs. That means their workers are paid less than workers in many other countries. In most cases, the large trainer companies do not own the factories or pay the workers directly. They use other companies who own the factories and employ the workers.

For several decades, there was not enough attention given to the way workers were paid and treated. Some factory owners even used children as cheap labour. However, over the past few years, international organisations, governments and consumers have encouraged the large trainer companies to start making their shoes in a way that is fairer to the workers. Today, **independent** organisations can visit the factories and check how the workers are being treated.

Most trainers are made in factories in these countries, and shipped to the rest of the world.

China

Bangladesh

Taiwan

Vietnam

India

Indonesia

Ethical Policies

It is now common for companies such as the large manufacturers of trainers to create statements and **policies** on the **ethics** of how they do business.

In these statements, they commit to:

- Treat workers fairly and equally and provide a safe working environment.
- Use natural resources responsibly and efficiently and reduce waste.
- Make sure their suppliers do not employ children to do work.
- Ensure products are safe and work the way they are supposed to.
- Source raw materials in ways that do not harm the environment.

Setting Goals

As well as releasing statements, some of the large companies that make trainers set themselves goals to help them stick to the promises in their ethical policies. These goals include increasing the use of environmentally preferred materials and reducing the amount of water used when the shoes are dyed.

Natural rubber is collected from a rubber tree.

Earning a Tick

It is hard for consumers to know if a company has met its stated ethical goals. There are resources available which rate companies and brands. These systems can help consumers make ethical choices when they shop.

STEP 4 Warehousing and Logistics

Warehousing is the storing of trainers in warehouses after they are made but before they are sent to shops or sold online. Logistics is the process of storing, moving and keeping track of the trainers as they are moved from factories to warehouses and then to shops.

Transportation

Although almost all trainers are made in Asia, Central America and South America, about two-thirds of trainers are sold in North America and Europe. That means they have to be transported a long way, often across oceans and continents. The high cost of transport can make trainers expensive to buy.

A container ship transports multiple layers of containers.

The most common forms of transport are ship and truck. Trainers are sometimes transported by plane, but only when they need to be sent quickly. It is much cheaper to transport goods by sea than by air.

In most years, about 78% of transportation of trainers is by ship, almost 20% by truck, and only 1–2% by plane.

A typical journey for a pair of trainers looks like this:

1. Trainers are transported by truck from the factory to a port.

2 Trainers are loaded onto a ship and transported by sea to the country where they will be stored and then sold.

3 Trainers are unloaded from the ship and transported by truck to huge warehouses in that country.

4 Trainers are transported by truck and van from the warehouses to retail stores all around the country.

Distribution Centres

After being transported, trainers are stored in distribution centres. Distribution centres are huge warehouses and are often located near ports, airports or major highways to make it easier to move goods to their next destination. Ideally, they do not store products for long. Companies want the products to be sent to stores as soon as possible after arriving at the distribution centre. This keeps the cost of storing them down.

Nike North America Logistics Campus

The Nike North America Logistics Campus in Tennessee, USA, is one of the largest distribution centres in the world. It is 260 000 square metres, and has more than 25 kilometres of **conveyor belts** that move boxes of trainers and other products within the centre.

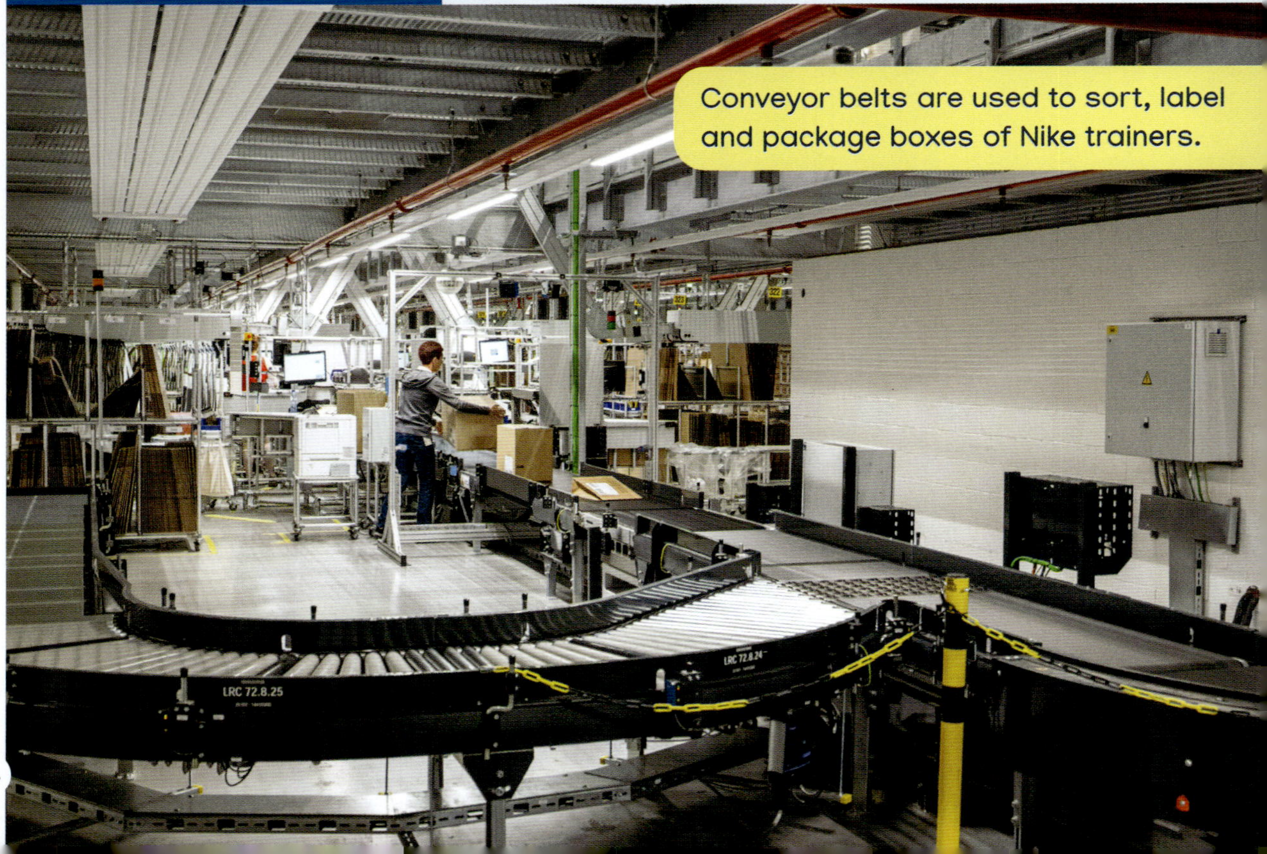

Conveyor belts are used to sort, label and package boxes of Nike trainers.

Carbon Footprint

The transportation of goods around the world is not good for the environment. It uses a great deal of fuel, which releases harmful gases such as carbon dioxide. How much carbon dioxide a process releases into the atmosphere is known as its "carbon footprint". Some of the companies that manufacture trainers try to reduce their carbon footprint in several ways.

Those companies choose to:

- Transport by ship rather than air, because air travel creates higher **carbon emissions**.
- Transport as many goods as possible in each shipment, because fewer shipments mean fewer carbon emissions.
- Use electric vehicles where possible for road deliveries.
- Use as much renewable energy as possible at factories and warehouses.

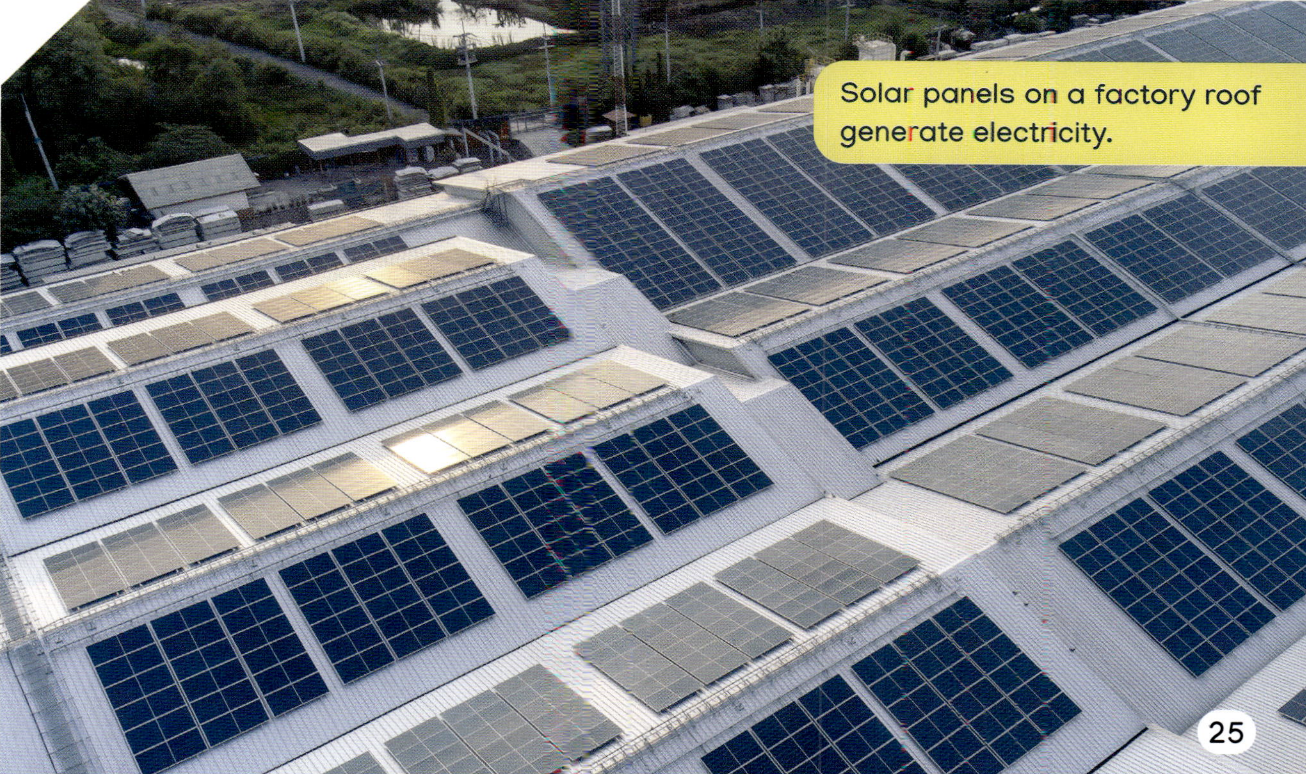

Solar panels on a factory roof generate electricity.

STEP 5 Retail

Have you and your family ever walked out of a retail store without buying something? Of course you have. Imagine if everyone who walked into a shop did the same thing. Nothing would be sold. All that effort in designing a product, collecting the raw materials, making the goods, transporting them around the world, storing them in warehouses and delivering them to the shop would have been wasted. So, in many ways, the retail step in the supply chain is the most important of all.

Selling Trainers

The companies that make trainers, and the stores that sell them, use different methods to try to convince consumers to buy the product.

Common methods of selling are to:

- Display the trainers so they can be seen clearly, such as in the window or on displays at eye level.
- Put them on sale, which is selling them at a cheaper price than normal.
- Explain to shoppers how particular trainers will improve their athletic performance or their health.
- Offer a fair return and exchange policy so that consumers know they can bring the trainers back if they need to.

Sales staff assist customers to select, size and try on shoes.

Online Retail

Trainers are not only sold in stores. They can also be bought online. There are positives and negatives to buying trainers online.

Online shopping can be a quick and easy way to buy goods.

- The trainers are delivered to your door.
- They may be cheaper, because an online seller's costs are generally lower than a retail store.
- It is easier to browse a lot of online stores than physical stores.

- You cannot try the trainers on to make sure they fit.
- They may be harder to return if there is something wrong with them.
- They can be damaged being transported.

Reusing Trainers

The supply chain does not always end when a pair of trainers is sold. When someone has outworn or outgrown their trainers, the trainers can be returned to the manufacturer to be recycled or donated to people in need. The recycled parts, such as polyester, can then become part of the early stages of the supply chain again.

Marketing and Advertising

The companies that make trainers spend more than $5 billion a year on marketing and advertising. That might seem like a lot, but consumers spend more than $60 billion a year on trainers.

Advertisements appear on television, websites, radio, newspapers, magazines, billboards, social media and in cinemas. Marketing includes paying sports stars to wear the company logos and names on their shirts when they play, paying social media influencers to mention the company's products, and paying celebrities to use and wear their products.

The most famous celebrity to use a brand of trainers is the basketball player Michael Jordan, who promotes the Nike Air Jordan trainers. Other celebrities who have promoted trainers include Justin Timberlake (Nike), Selena Gomez (Puma) and Lionel Messi (adidas).

Michael Jordan holds an Air Jordan XX3 trainer.

A Global Product

The supply chain for trainers involves thousands of people, spread across thousands of kilometres, in many countries and different continents. It's possible that the trainers you wear have connections to four or five continents. They may have been designed in North America or Europe, contain raw materials from Africa, manufactured in Asia and sold in Australia. Trainers truly are global products.

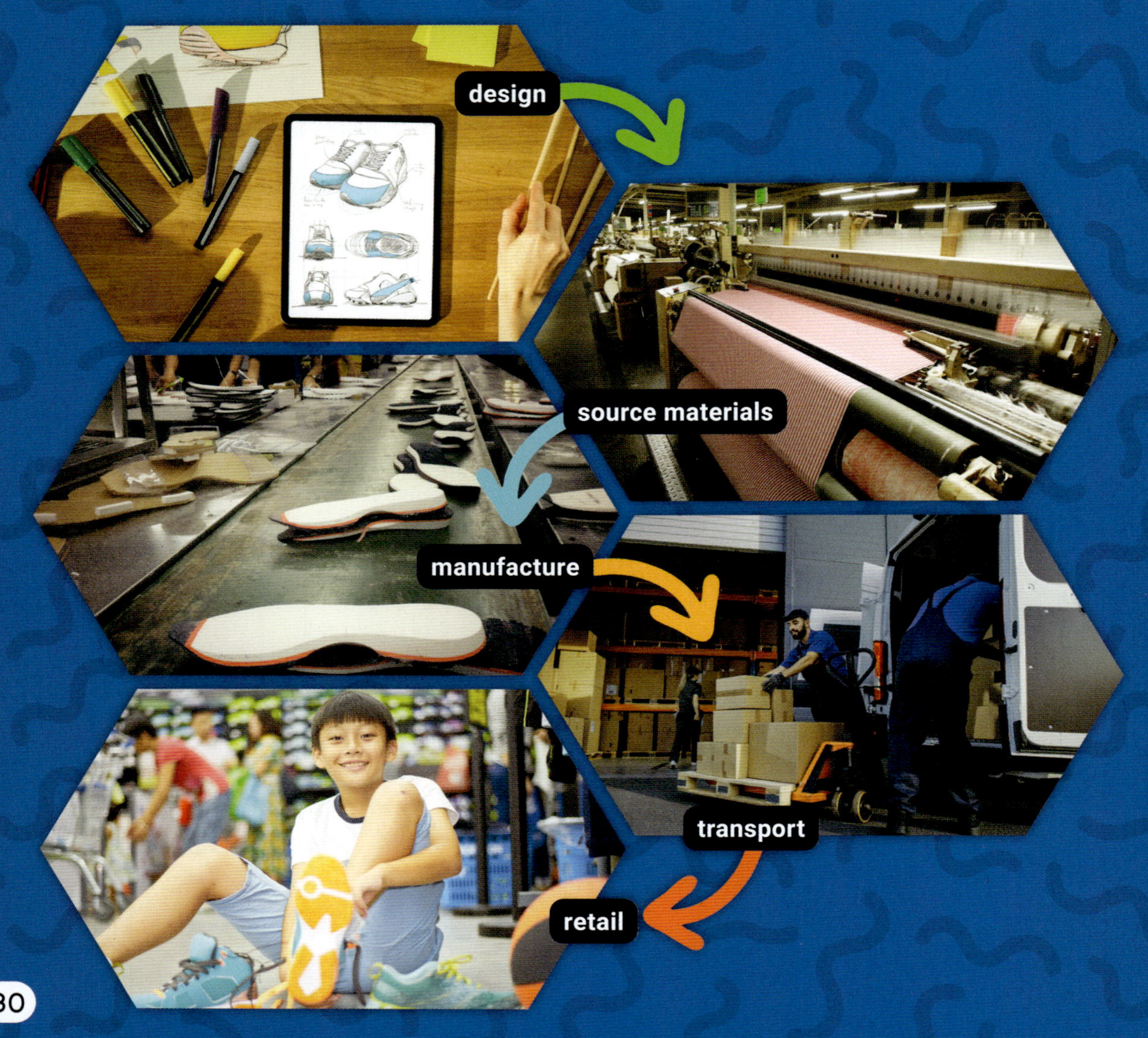

Glossary

artificial leather (*noun*)	a material made from layers of foam, plastic and a top layer which looks like leather
carbon emissions (*noun*)	the gases released into the atmosphere by burning fossil fuels
colonisation (*noun*)	when one country takes control of another
consumers (*noun*)	people who buy or use products
conveyor belts (*noun*)	continuously moving bands that transport goods within a warehouse or factory
ethics (*noun*)	behaviour or actions made in a way that is good for others
features (*noun*)	special or important details about a thing
fibres (*noun*)	materials made up of many threads
goods (*noun*)	items that are bought and sold
independent (*adjective*)	not controlled by someone else
industrial design (*noun*)	designing things used by people and businesses
initiative (*noun*)	an action or step to change something
labour (*noun*)	work or workforce
logistics (*noun*)	the organisation of transporting goods
manufacture (*verb*)	to make something
manufacturers (*noun*)	the companies that make products, usually in a factory
policies (*noun*)	official records on how things are to be done
port (*noun*)	a place where ships land and leave from
practices (*noun*)	the ways in which something is done
raw materials (*noun*)	the basic parts that something is made from
retail (*noun*)	the selling of goods to the public
simulates (*verb*)	re-creates or copies
thrive (*verb*)	to succeed and remain successful

Index

Earthquakes

Julie Haydon

Australia • Brazil • Japan • Korea • Mexico • Singapore • Spain • United Kingdom • United States

Earthquakes

Fast Forward
Gold Level 22

Text: Julie Haydon
Illustrations: Boris Silvestri
Editor: Cameron Macintosh
Design: Stella Vassiliou
Series design: James Lowe
Production controller: Seona Galbally
Photo research: Fiona Smith
Audio recordings: Juliet Hill, Picture Start
Spoken by: Matthew King and Abbe Holmes

Acknowledgements
The author and publisher would like to acknowledge permission to reproduce material from the following sources: Cover; Alamy Images/Pacific Press Service/Yoshiaki Nagashima; AP Photo/Steven Wang, p14; Alamy Images/Pacific Press Service/Yoshiaki Nagashima, p 1; Corbis/Bettmann, p 5, 13/ Shahpari Sohaie, pp 18-19; Getty Images/John Barr, p4 right/ AFP, p16/ Majid/Stringer, p19/ Majid, p20/ AFP/Jean-Francois Camp, p21 top/ Majid, p21 bottom/ AFP/Martin Bureau, p22/ Scott Peterson, p23; iStockphoto, p 3, 12; Photolibrary, pp 10, 18 top left, Newspix/AFP Photo/Yoshikazu, p17.

ISBN 978 0 17 012686 1
ISBN 978 0 17 012681 6 (set)

Cengage Learning Australia
Level 7, 80 Dorcas Street
South Melbourne, Victoria Australia 3205
Phone: 1300 790 853

Cengage Learning New Zealand
Unit 4B Rosedale Office Park
331 Rosedale Road, Albany, North Shore NZ 0632
Phone: 0508 635 766

For learning solutions, visit cengage.com.au

Printed in Australia by Ligare Pty Ltd
9 10 11 12 13 14 15 20 19 18 17 16

THE UNIVERSITY OF MELBOURNE

Evaluated in independent research by staff from the Department of Language, Literacy and Arts Education at the University of Melbourne.

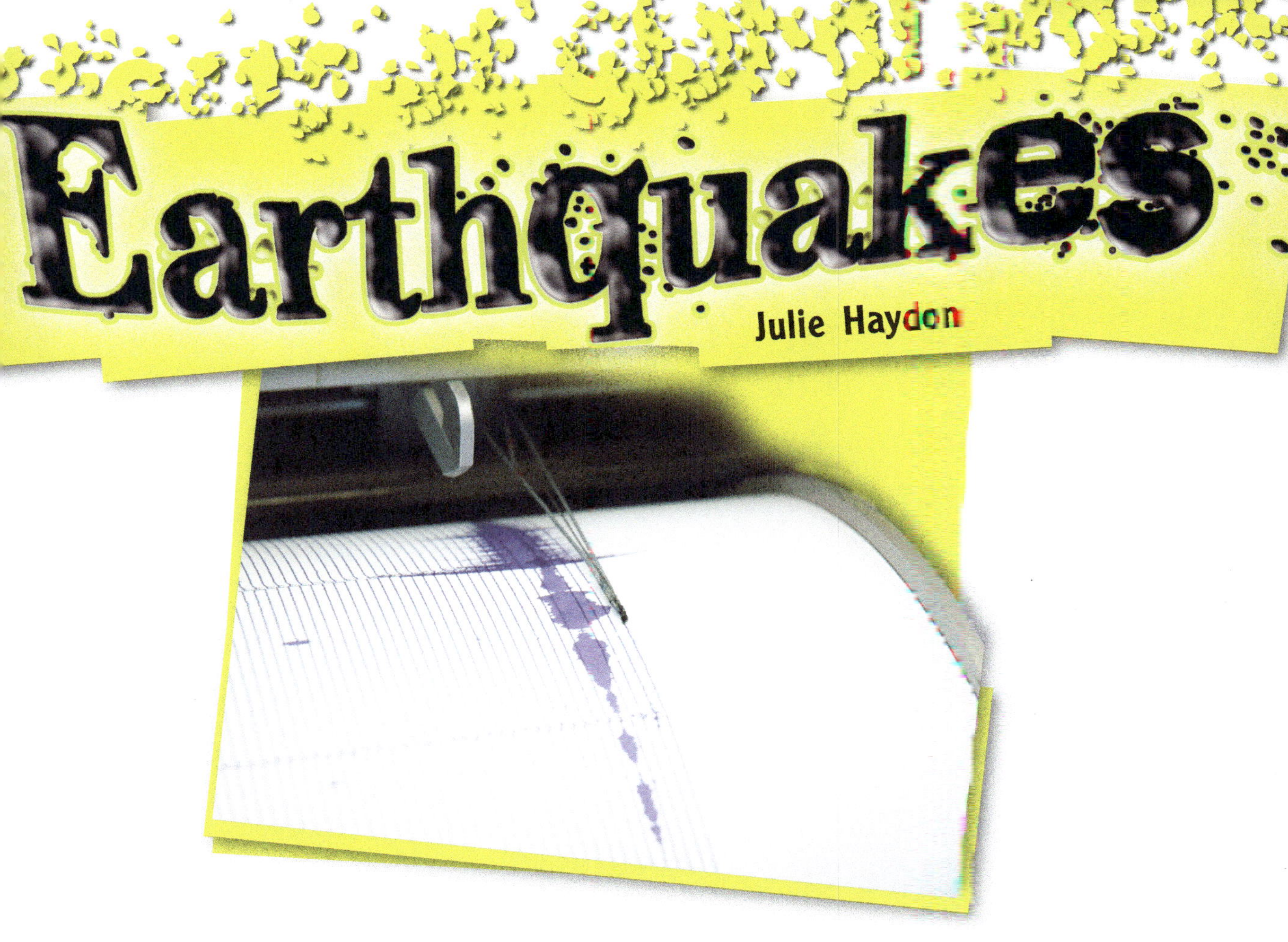

Contents

WHAT IS AN EARTHQUAKE?

An earthquake is the shaking of the Earth's surface. Earthquakes are set off by fast movements of the Earth's rocky shell.

There are about 500 000 **detectable** earthquakes every year.

They occur in many places around the world, including under the oceans.

Most earthquakes are small and don't cause any damage, but large earthquakes can be devastating.

Mexico City, Mexico, 1985

Scientists study the Earth's layers to understand earthquakes. Recently, scientists have been able to explain how and why earthquakes happen.

Tangshan, China, 1976

The science of earthquakes is called seismology. Scientists who study earthquakes are called seismologists.

Chapter 2

THE EARTH'S LAYERS

Long ago, people believed that the Earth was solid, but today, scientists know that the Earth is made up of layers.

The centre of the Earth is called the core.
The core is very hot.
The inner core is a solid ball made mostly of iron, and the outer core is made mostly of liquid iron.

The mantle surrounds the core and is made of very hot rock.
The lower part of the mantle is soft and can move very slowly.
The mantle and the core make up most of the Earth.

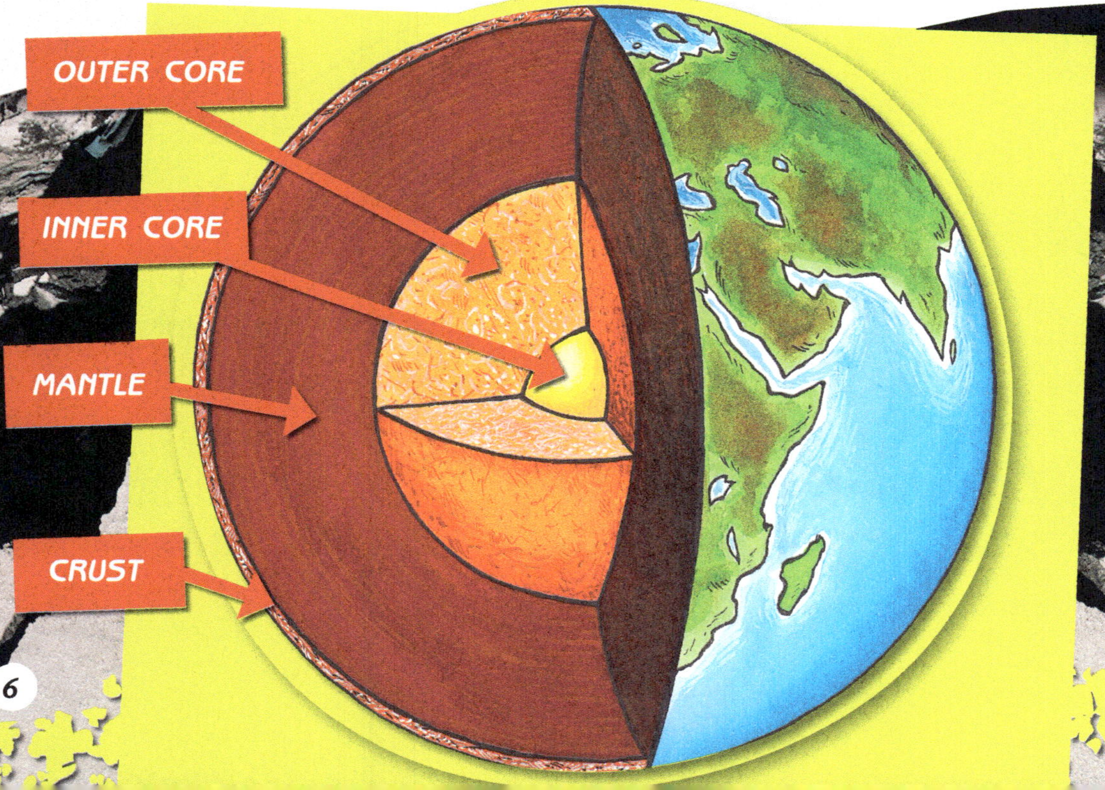

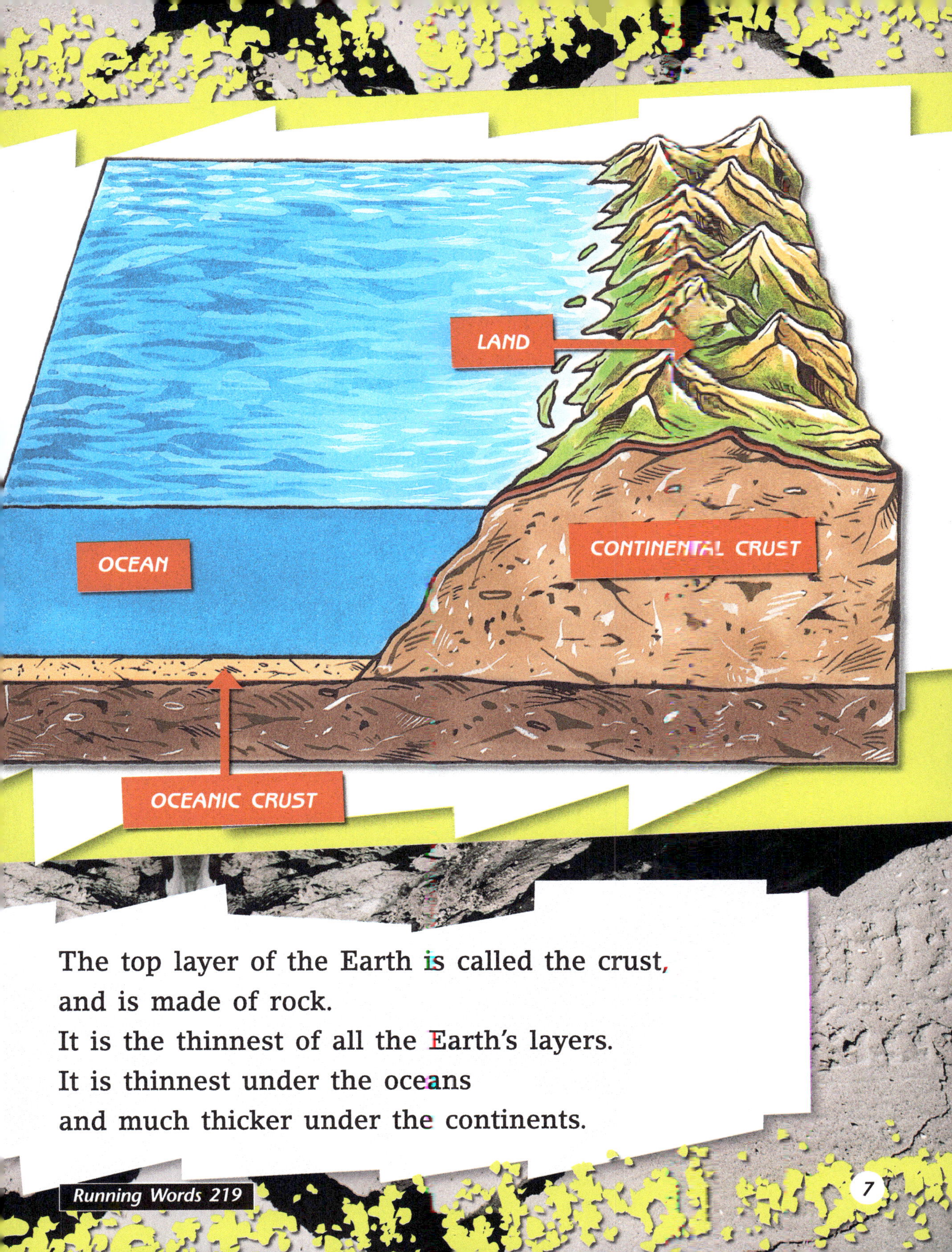

The top layer of the Earth is called the crust,
and is made of rock.
It is the thinnest of all the Earth's layers.
It is thinnest under the oceans
and much thicker under the continents.

TECTONIC PLATES

The crust and the upper part of the mantle are made up of huge, plate-like pieces of rock that float on the lower part of the mantle, so they are always moving.

Most of the time, the movements of the pieces are so slow that people cannot feel them.

These huge pieces of rock are called tectonic plates.

They are different sizes, and each one has a name.

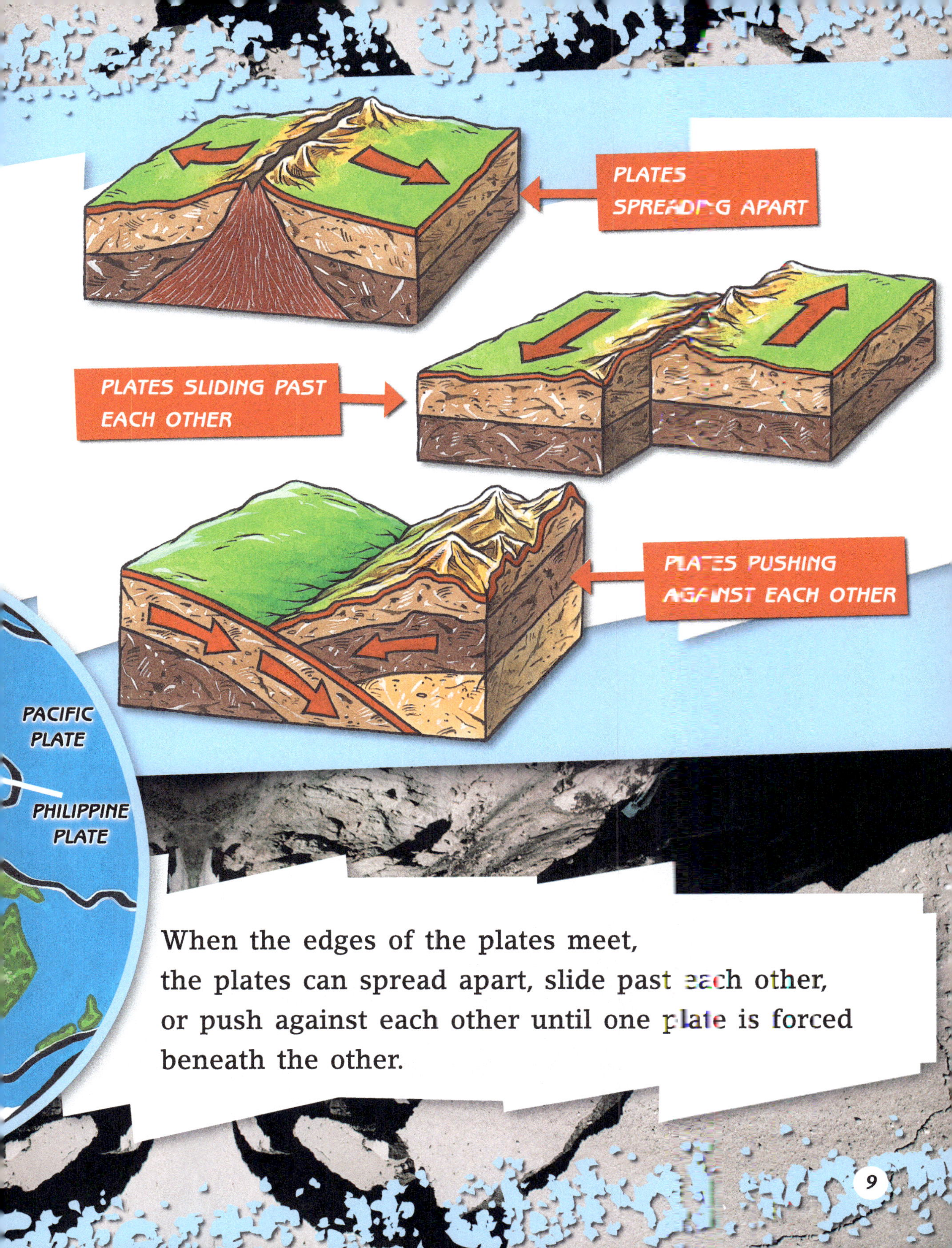

When the edges of the plates meet,
the plates can spread apart, slide past each other,
or push against each other until one plate is forced
beneath the other.

FAULTS

The movements of the tectonic plates make cracks in the Earth's rocky crust.
These cracks are called faults,
and they form at the edges of the plates.
Some faults are easy to see.

San Andreas Fault, California, USA

Most faults are stuck in place,
but the plates keep moving slowly.
Over time, this builds stress in the rocks.
When the stress is too much for the rocks,
they break suddenly along the fault.
Shock waves spread out in all directions,
and make the ground shake.
This is an earthquake, and it is why most earthquakes
happen near the edges of tectonic plates.

Chapter 5

MEASURING EARTHQUAKES

Scientists use machines called seismographs to measure earthquakes.
Information from seismographs helps scientists to work out the size of an earthquake, and where and when it happened.
A small earthquake makes a small wiggle on a seismograph, and a big earthquake makes a large wiggle.

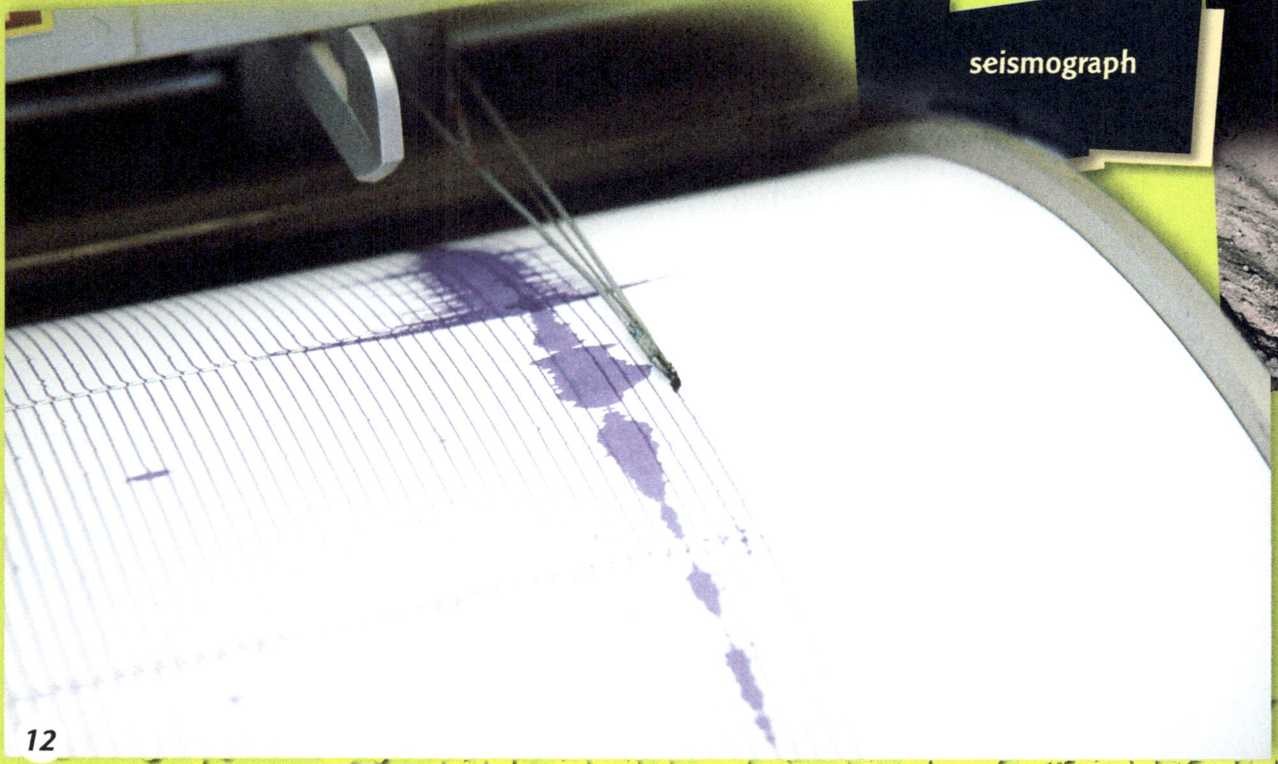

seismograph

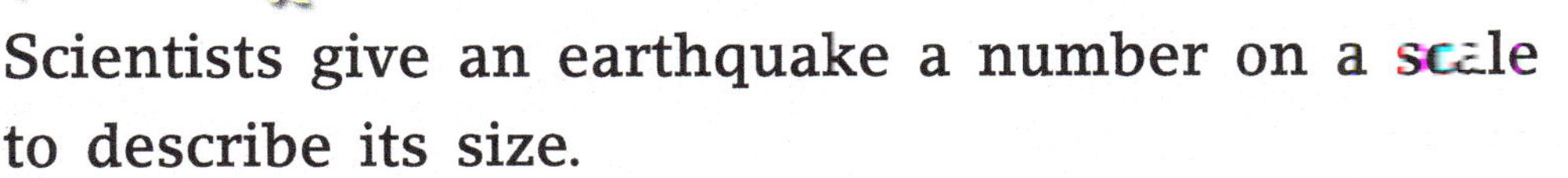

Scientists give an earthquake a number on a scale to describe its size.
The higher the number, the bigger the earthquake.
There are several different scales, but scientists often use the **Magnitude** scale to measure the size of the earthquake source.
This magnitude 9.5 earthquake in Chile in 1960 was the world's biggest earthquake since 1900.

Valdivia, Chile, 1960

The Richter scale measures the largest wiggle on the recordings made by the seismographs.

The Modified Mercalli scale measures the damage the earthquake does on the Earth's surface. An earthquake that is high on the Richter scale may not do a lot of damage in an area where few people live. The same earthquake would be low on the Modified Mercalli scale.

Modified Mercalli Scale

Average peak velocity	Intensity value and description	Average peak acceleration
	I. Felt only by a very few people.	
	II. Felt only by a few people at rest, especially on upper floors of buildings.	
	III. Felt by people indoors, especially on upper floors of buildings but many people do not recognise it as an earthquake. Standing cars may rock slightly. Vibrations similar to the passing of a truck.	
1–2	IV. Felt indoors by many, outdoors by few during the day. At night, some people awakened. Dishes, windows and doors disturbed; walls creak. Sensation like a heavy truck striking a building. Standing cars rock noticeably.	0.015g–0.02g
2–5	V. Felt by nearly everyone; many awakened. Some dishes and windows broken.	0.03g–0.04g
5–8	VI. Felt by all, many frightened. Some heavy furniture moved. Damage slight.	0.06g–0.07g
8–12	VII. Everybody runs outdoors. Not much damage in buildings of good design and construction; lots of damage in poorly built or badly designed structures. Some chimneys broken.	0.10g–0.15g
20–30	VIII. Damage slight in specially designed structures. Lots of damage in many ordinary buildings. Damage great in poorly built structures. Chimneys, monuments and walls fall.	0.25g–0.30g
45–55	IX. Lots of damage in specially designed structures; damage great in many buildings. Cracks in ground.	0.50g–0.55g
more than 60	X. Many structures destroyed. Rails bent.	more than 0.60g
	XI. Few structures remain standing. Bridges destroyed. Large holes in the ground. Rails bent greatly.	
	XII. Damage total. The ground moves in waves. Objects thrown into the air.	

Chapter 6

PREPARING FOR AN EARTHQUAKE

Scientists don't know exactly when an earthquake will happen, but they do know that earthquakes are more likely to happen in some areas. People living in these areas have to be prepared for an earthquake.

Office workers in South Korea practise their escape in preparation for future earthquakes.

Earthquakes are common in Japan.
During an earthquake, school children sit under a table so that they are protected from falling objects.
If they are outside, they try to move to an open area.
Some people keep a helmet handy in case of an earthquake.

Chapter 7

EARTHQUAKE SURVIVOR

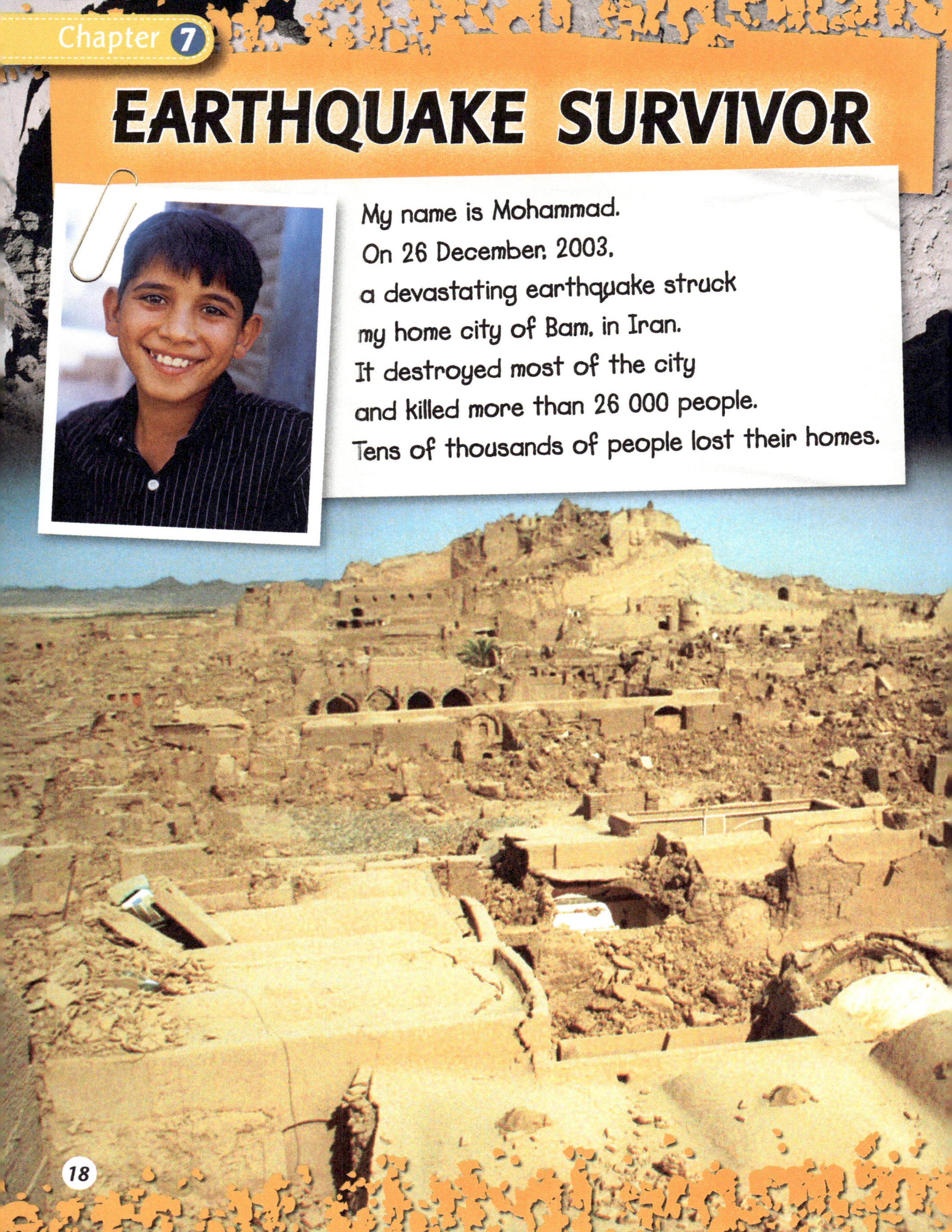

My name is Mohammad.
On 26 December, 2003,
a devastating earthquake struck
my home city of Bam, in Iran.
It destroyed most of the city
and killed more than 26 000 people.
Tens of thousands of people lost their homes.

It was early morning when the earthquake struck.
I was asleep with my family outside our home.
A smaller earthquake had woken us earlier in the night so we decided to sleep outside.
This decision saved our lives.

The earthquake made the ground shake.
It seemed to rise in waves beneath us.
I was very scared.

Most of the houses were made of mud bricks. They collapsed, and people were buried beneath them. All over the city, people helped to dig out survivors. My father and I helped to rescue some of our neighbours.

Bam, before the earthquake

Bam is over 2000 years old,
and was very famous for its history and beauty.
The magnitude 6.6 earthquake turned most of the city into a pile of dust.

Bam, after the earthquake

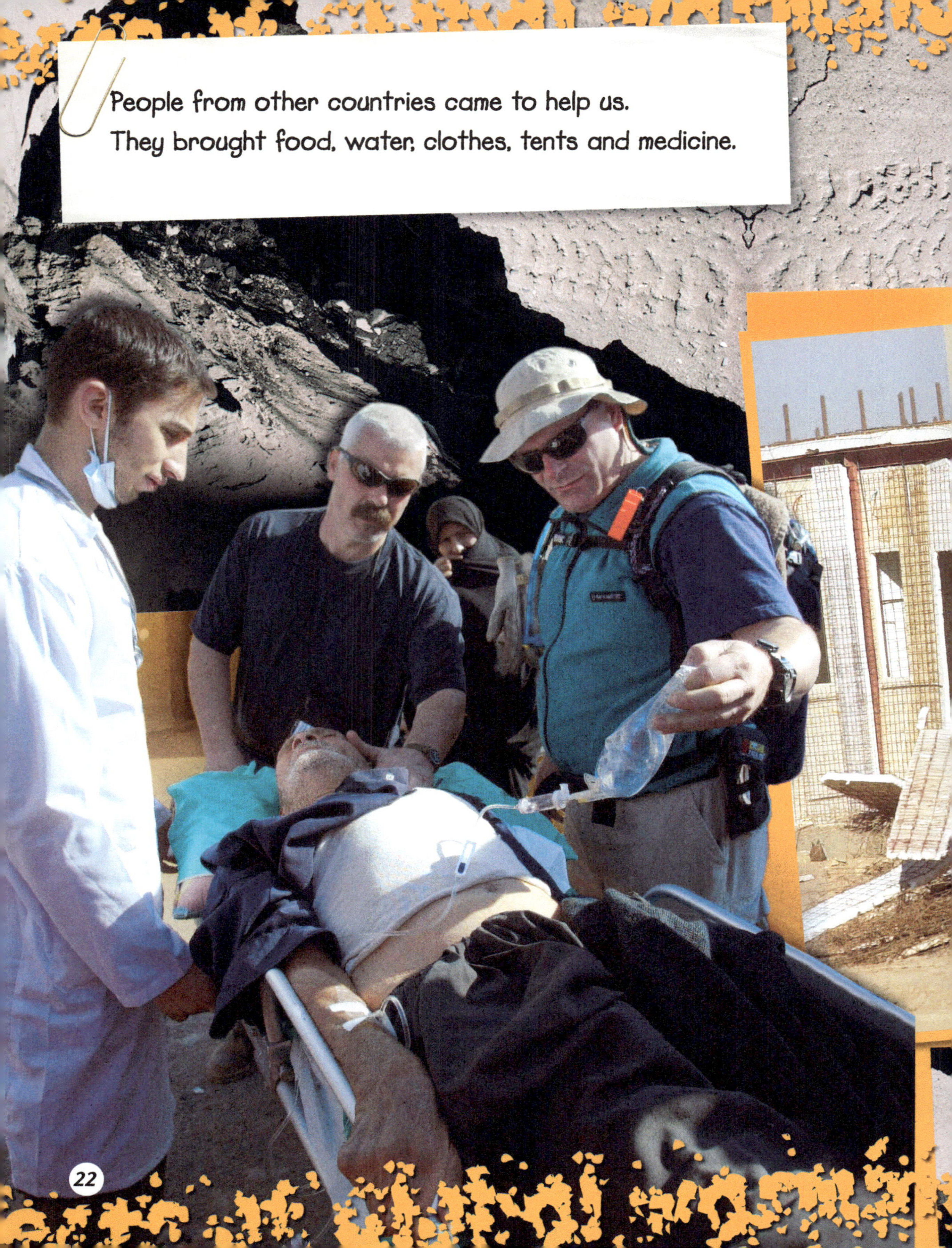

People from other countries came to help us.
They brought food, water, clothes, tents and medicine.

People have started to build new homes in Bam, but the city will never be the same.

Glossary

detectable able to be felt, or detected by scientific instruments

magnitude the size of something

Index

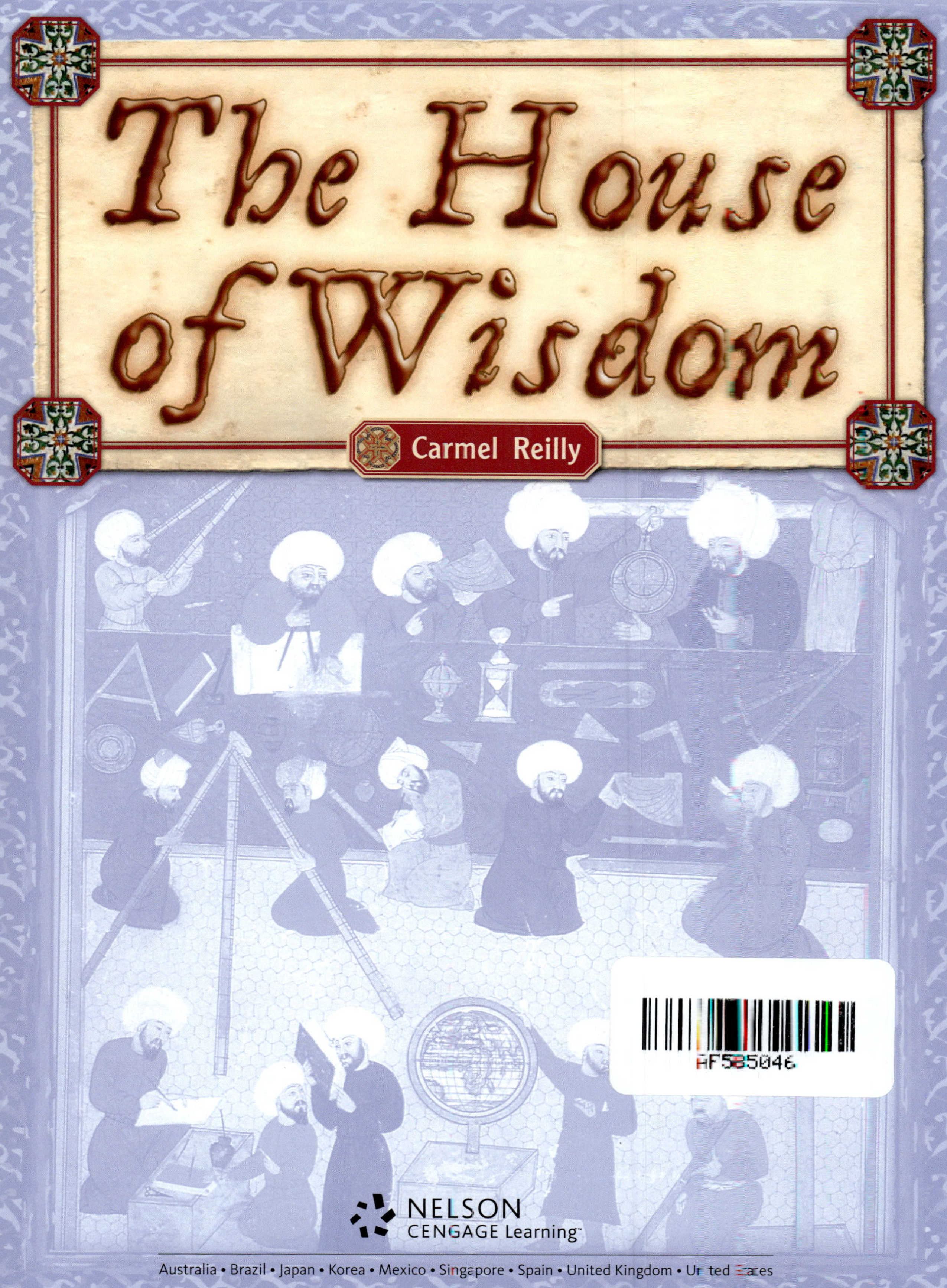
The House of Wisdom
Carmel Reilly
AF585046
NELSON
CENGAGE Learning
Australia • Brazil • Japan • Korea • Mexico • Singapore • Spain • United Kingdom • United States

The House of Wisdom

Fast Forward
Silver Level 24

Text: Carmel Reilly
Illustrations: Boris Silvestri
Editor: Johanna Rohan
Design: Stella Vassiliou
Series design: James Lowe
Production controller: Seona Galbally
Photo research: Gillian Cardinal
Audio recordings: Juliet Hill, Picture Start
Spoken by: Matthew King and Abbe Holmes
Reprint: Jennifer Foo

Acknowledgements
The author and publisher would like to acknowledge permission to reproduce material from the following sources: Front & back covers: Photolibrary/akg-images, London, p 18; The Art Archive/British Library, p 23/Bodleian Library Oxford, p 5/ Dagli Orti, p 20/ Edinburgh University, p 22/Galleria degli Uffizi Florence/Dagli Orti, p 4 left and right; Photolibrary, pp 1, 8, 10, 13, 16-17.

ISBN 978 0 17 012714 1
ISBN 978 0 17 012705 9 (set)

Cengage Learning Australia
Level 7, 80 Dorcas Street
South Melbourne, Victoria Australia 3205
Phone: 1300 790 853

Cengage Learning New Zealand
Unit 4B Rosedale Office Park
331 Rosedale Road, Albany, North Shore NZ 0632
Phone: 0508 635 766

For learning solutions, visit cengage.com.au

Printed in Australia by Ligare Pty Ltd
5 6 7 8 9 10 11 21 20 19 18 17

THE UNIVERSITY OF MELBOURNE

Evaluated in independent research by staff from the Department of Language, Literacy and Arts Education at the University of Melbourne.

Contents

INTRODUCTION

Modern science began about 600 years ago when a number of important discoveries were made in the areas of mathematics and **astronomy**.

In the 1400s and 1500s, great scientific thinkers such as Copernicus and Galileo changed the way that people looked at the world. They challenged the idea that the Earth was at the centre of the universe. Then, over the next few hundred years, other scientists started to make other breakthroughs, which led to the beginnings of industry, technology and modern medicine.

Copernicus

Galileo

Euclid, (left) a mathematician in ancient times

However, many of these breakthroughs had their beginnings much earlier. For thousands of years, people had been observing the world around them and testing out mathematical and scientific ideas. In ancient times, the Egyptians, Romans and Greeks had all made important advances in science. During the **Middle Ages**, many important advances in science were made in the Islamic world.

Chapter 2

THE HOUSE OF WISDOM

In the 600s AD, the religion and culture of **Islam** began to grow in the Middle East. Within 200 years, the Islamic Empire had spread from its capital in Baghdad to Turkey, North Africa and Spain.

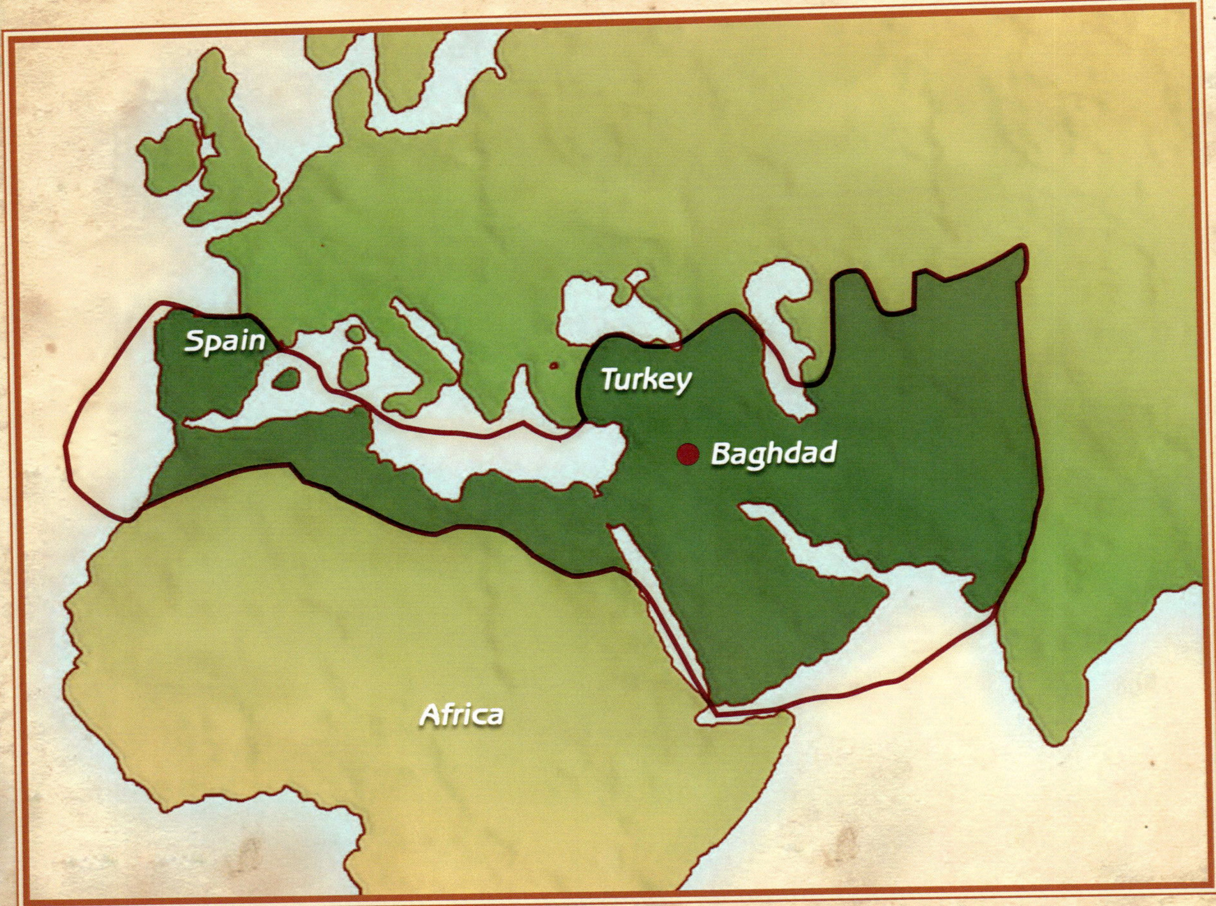

the Islamic Empire, around 750 AD

In the early 800s AD, the Islamic Empire was ruled by a **caliph** called al-Mamun. Al-Mamun believed that culture and learning were important. He understood that cultures needed to build on discoveries from the past so that they could grow. He also thought it was important to create places where people could work, study and pass on knowledge. Because of this, al-Mamun decided to build a large library and place of learning in Baghdad. He called this place the House of Wisdom *(Bayt al–Hikmah)*.

Running Words 262

The House of Wisdom became a centre of learning in many areas, but it was most important in the areas of mathematics and science.

Most of the books housed at the House of Wisdom were books from ancient Greek times and before. They covered topics such as philosophy, astronomy, medicine and chemistry. In some cases they were the only copies of these books anywhere in the world at that time.

a painting of Socrates, an ancient Greek philospher, by a Middle Eastern artist around the time of the House of Wisdom

The House of Wisdom also held important ancient maps, models and works of art. Because of its treasures, people came from all over the Islamic Empire to teach and study there.

AL-KHWARIZMI

One of the most famous scholars at the House of Wisdom was a man called al-Khwarizmi, who was a mathematician and astronomer. Al-Khwarizmi spent much of his time at the House of Wisdom studying and translating Greek scientific books into Arabic. He was famous for introducing the number system we use today to the world.

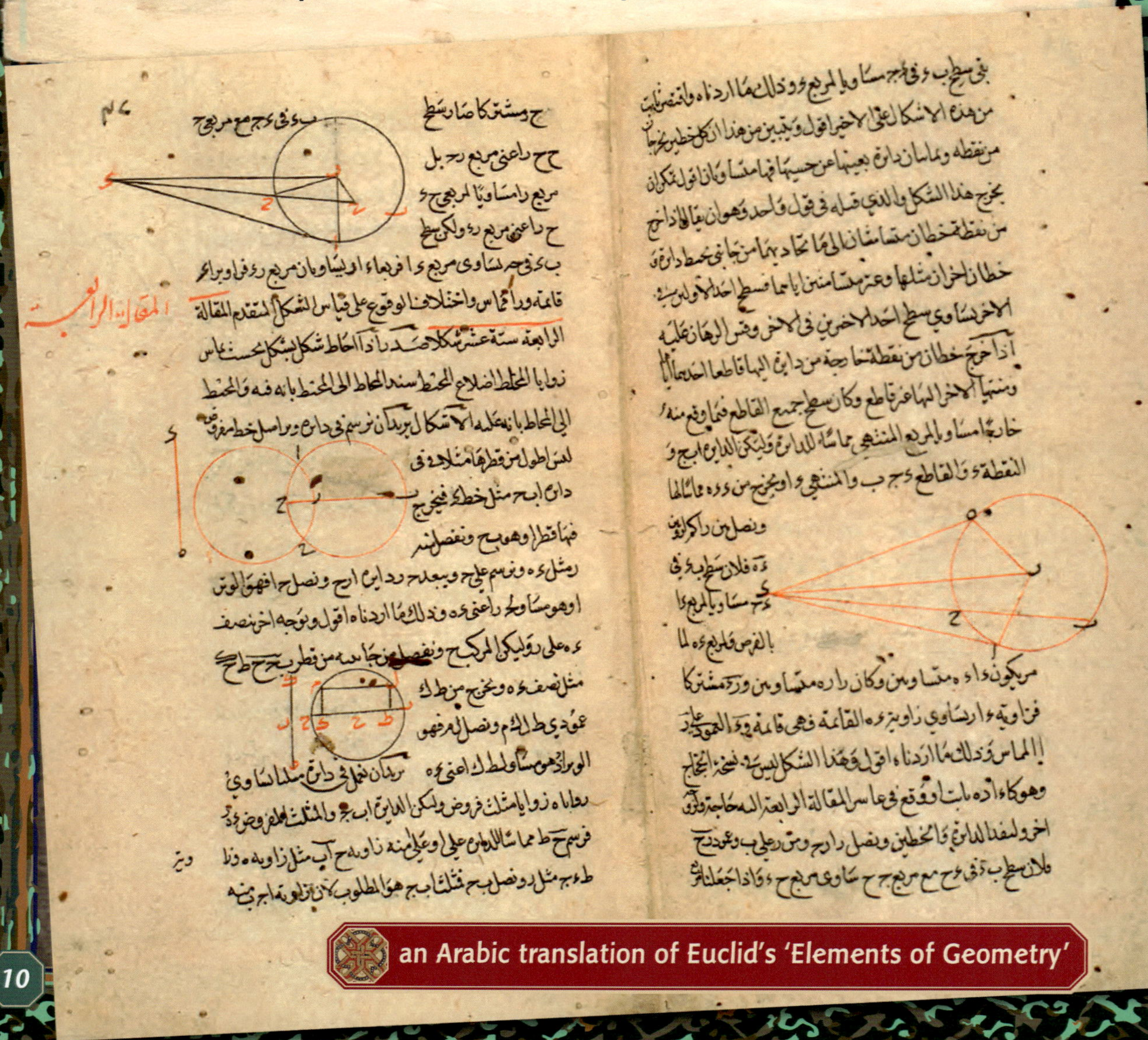

an Arabic translation of Euclid's 'Elements of Geometry'

Numbers

The number system of 0, 1, 2, 3, etc. originally came from India. When al-Khwarizmi first saw it, he realised that it was much easier than using other systems, such as **Roman numbers**. He also took the Indian idea of using zero as a place-holding number, and counting in units of ten. He then introduced this system to the Islamic world, and over time it was passed on to the rest of the world.

Algebra

Al-Khwarizmi is known as the "father of algebra". Algebra is a mathematical system that uses letters and other symbols to represent numbers in order to solve mathematical problems. The name algebra comes from an Arabic word, *al-gabr*, which makes up part of the title of a book al-Khwarizmi wrote on this kind of mathematics, based on ancient Greek ideas.

Sir Isaac Newton

One thousand years later, Sir Isaac Newton used al-Khwarizmi's algebra to develop the basis of calculus. Calculus is a special way of understanding practical problems by using algebra, and Newton was able to use this to work out the rules of gravity. Later, scientists in the 20th century went on to use Newton's calculus to work out other things, such as how to launch rockets into space.

AL-KINDI

Al-Kindi was another important scholar in the early years of the House of Wisdom. He was a philosopher, scientist, mathematician and medical doctor, and he wrote more than 200 books on these subjects.

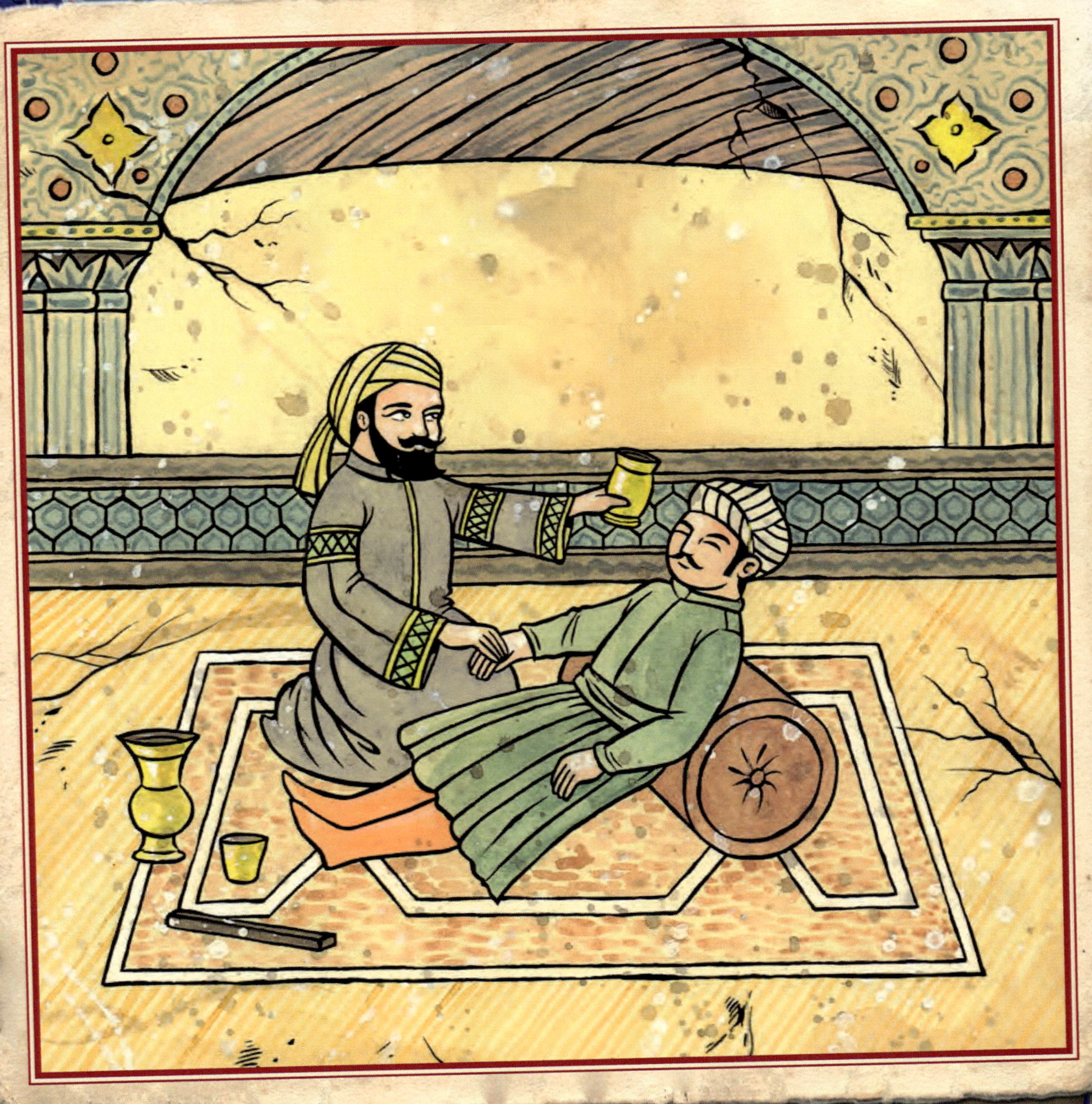

As a doctor, al-Kindi discovered how to give the right doses of medicine to people when they were ill. His work in chemistry also led him to understand a lot about metals and their properties.

THE BANU MUSA BROTHERS

The three Banu Musa brothers were also important at the House of Wisdom. They developed many new mathematical ideas that were eventually passed onto the rest of the world. These include using numbers, rather than explanations, to show areas and angles. This is called geometry.

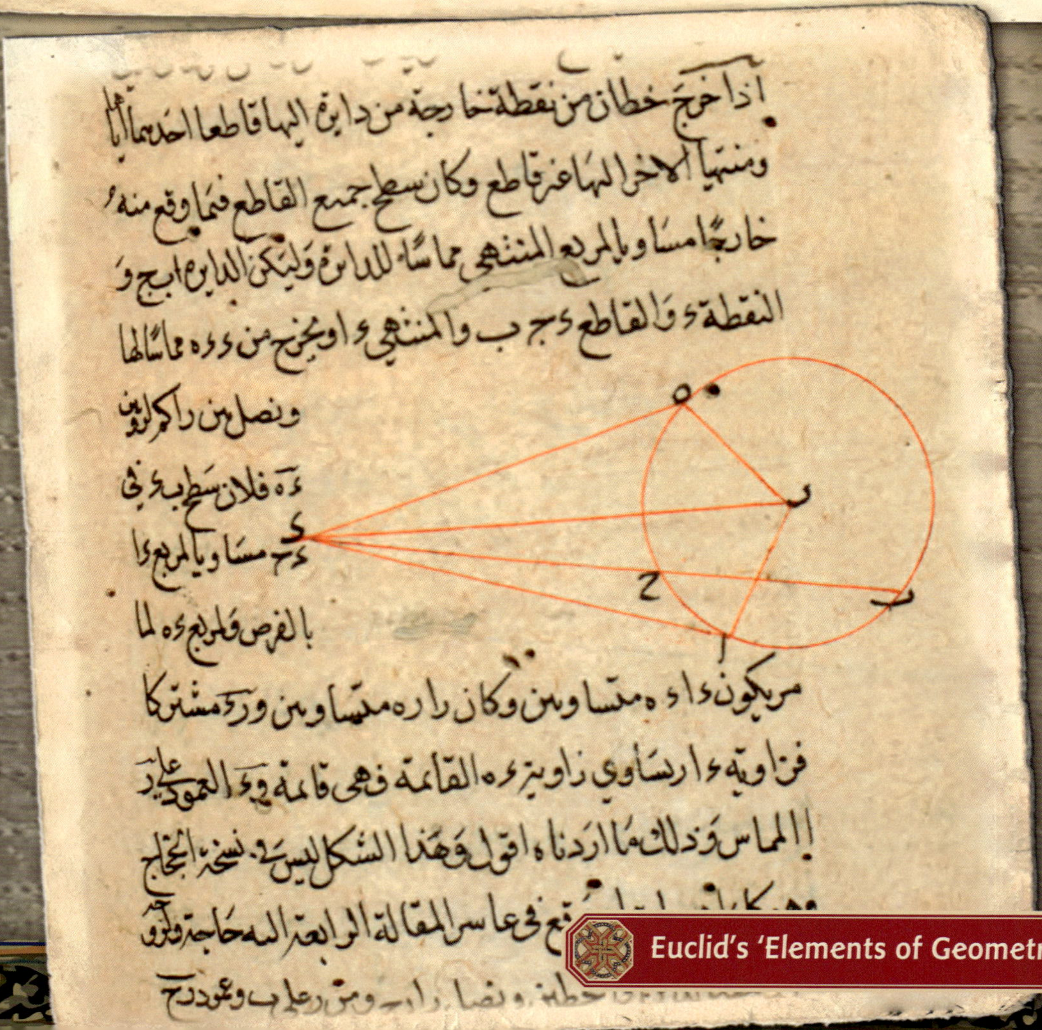

Euclid's 'Elements of Geometry'

The Banu Musa brothers also made many discoveries in astronomy, including working out that it took 365.6 days for the Earth to travel around the Sun.

Chapter 6

A PRACTICAL PLACE

The work that was done at the House of Wisdom was always seen as being practical and useful.

Medicine

Many of the discoveries made at the House of Wisdom were to do with medicine. A well-known scholar – Husayn ibn Sina – translated many medical books from Greek. He was also famous for his study of the eye. He wrote books that included detailed drawings of the eye, as well as other books on the human body. These books were used for centuries in both Islamic and European universities to teach medical students.

Engineering

Using the mathematics developed at the House of Wisdom, Islamic engineers were able to design and build many things. They made the perfect waterwheel, and constructed underground water channels that could supply water over long distances.

wooden waterwheels

Agriculture

Many Islamic scholars studied soils, weather and water. They were then able to advise farmers on which types of crops would grow best in certain places. This was important as the Islamic Empire was growing and so was its need for good, reliable sources of food for all of its people.

Chapter 7

THE END OF THE HOUSE OF WISDOM

The House of Wisdom remained one of the centres of Islamic learning for hundreds of years. However, it was completely destroyed when Baghdad was invaded by the **Mongols** in 1230 AD.

Mongol invaders

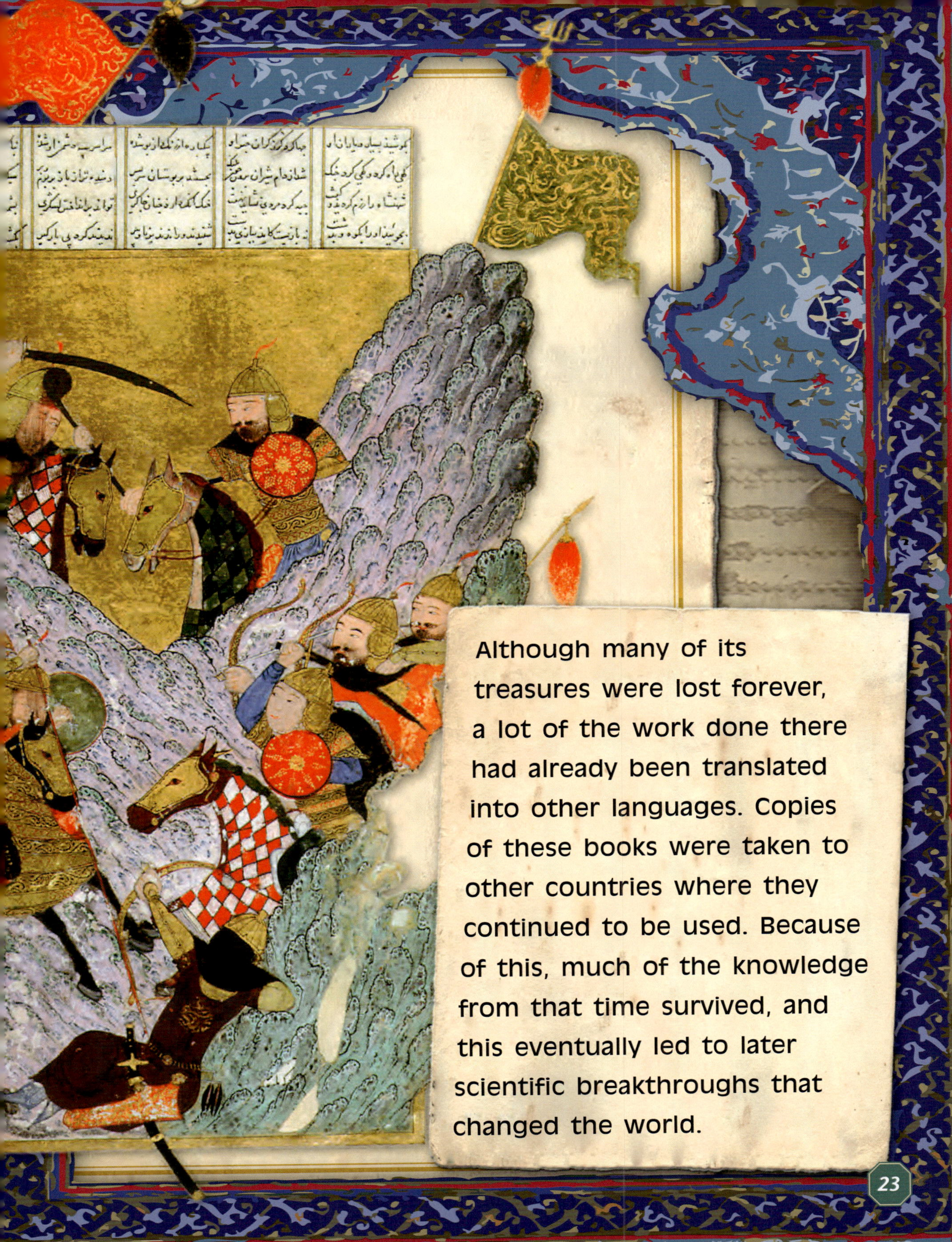

Although many of its treasures were lost forever, a lot of the work done there had already been translated into other languages. Copies of these books were taken to other countries where they continued to be used. Because of this, much of the knowledge from that time survived, and this eventually led to later scientific breakthroughs that changed the world.

Glossary

astronomy the science of stars, planets, comets, and galaxies

caliph the chief ruler

Islam the Muslim religion

Middle Ages the period in European history from the 5th century to the end of the 15th century. It was a period of great cultural, political and economic change.

Mongols native people of Mongolia

Roman numbers a numeral system originating in ancient Rome, using letters to represent numbers

Index